Canadian Pacific in the Rockies

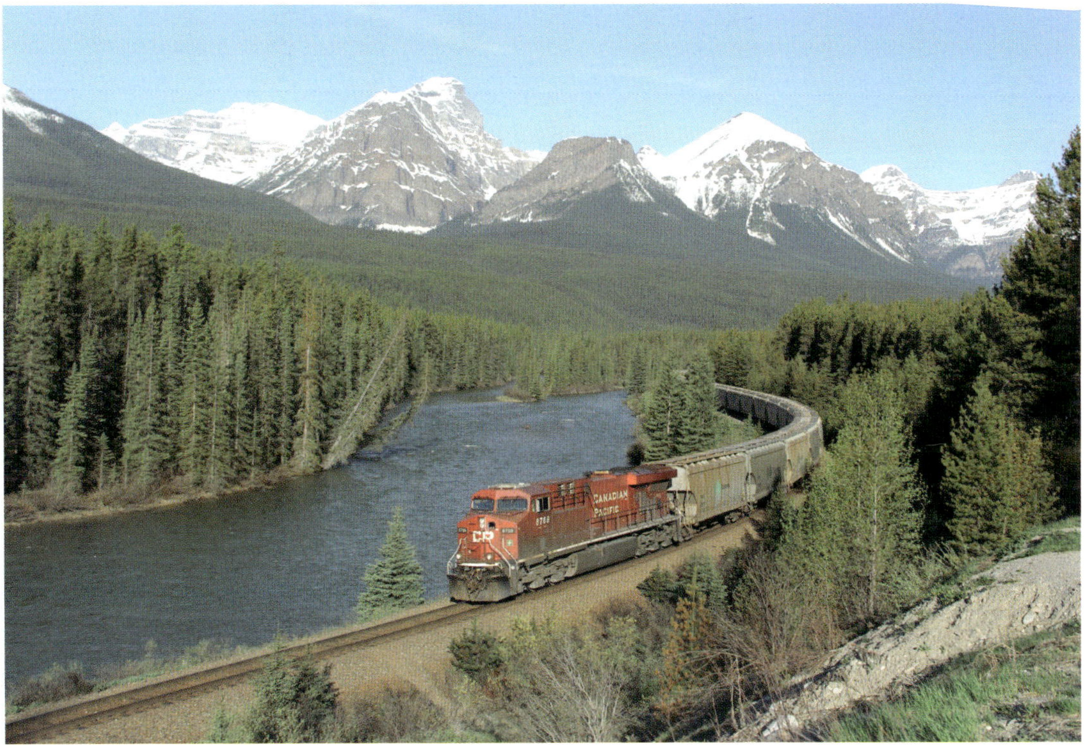

IAN LOTHIAN

KEY Books

Front cover image: On 14 October 2007, 9594 and 9156 were working hard on the approach to Banff East with a heavy westbound manifest freight from Calgary to Vancouver.

Back cover image: On 23 May 2017, the Royal Canadian Pacific ran from Calgary to Banff hauled by its three FP9 locomotives. With 1401 leading 4106 and 4107, the train is seen as it approaches Banff East Siding.

Title page image: On 28 May 2017, 8769 passed Morant's Curve leading an eastbound empty potash train that had 9376 as a mid-train helper.

Contents page image: A train of loaded tanks on their way to Vancouver passes Ottertail on 26 May 2017 behind CP 8723 and 8541.

Published by Key Books
An imprint of Key Publishing Ltd
PO Box 100
Stamford
Lincs PE19 1XQ

www.keypublishing.com

The right of Ian Lothian to be identified as the author of this book has been asserted in accordance with the Copyright, Designs and Patents Act 1988 Sections 77 and 78.

Copyright © Ian Lothian, 2021

ISBN 978 1 80282 001 0

All rights reserved. Reproduction in whole or in part in any form whatsoever or by any means is strictly prohibited without the prior permission of the Publisher.

Typeset by SJmagic DESIGN SERVICES, India.

Contents

Introduction 4
Chapter 1 Exshaw to Canmore 6
Chapter 2 Canmore to Banff 19
Chapter 3 Banff to Divide 39
Chapter 4 Divide to Field 55
Chapter 5 Field to Golden 67
Chapter 6 Cowley to Blairmore 79
Chapter 7 Blairmore to Sparwood 88

Introduction

Why do I have a fascination with the Canadian Pacific Railway is a question that I have often been asked so I will try to offer an explanation. When I was at high school, the class was told one day that for homework we had to write an essay on where we would like to visit most of all in the world. I chose to write about the Canadian Rockies and they have fascinated me for most of my life. Being brought up living close to a railway line, I grew up as a railway enthusiast who had a love of the Rockies and later in my life, I added photography as another hobby.

In all interests, there are always people whose work and achievements one admires and one photographer whose work I hold in very high esteem is Nicolas Morant, once the publicity photographer for Canadian Pacific. Looking at his photographs, the scene was set for my interest in the Canadian Pacific Railway to germinate. It would be many years later before I was able to follow the tracks west from Exshaw, past Canmore and Banff to Morant's Curve. There I was able to stand and understand why that location was a favourite of Nicolas Morant as the views there are spectacular.

It was then onward to Lake Louise and up and over the Continental Divide before encountering Kicking Horse Pass, the 'Big Hill' to generations of railwaymen, before the last leg of the journey to Golden. People can develop many addictions but I have come to realise that the Rockies are addictive; when I have been there, the time to leave is the time to plan my next visit. I can truthfully say that I know the area as well, if not better, than many parts of my own country and I hope that this collection of my photographs might convey the pleasure I have had on all of my visits.

I have to thank the friends that I have made along the way who have all helped me to photograph some unusual and interesting trains in so many locations. If anyone wonders what right someone from Scotland has to produce a book on the Canadian Pacific Railway (CPR) then they do not know the history of the CPR. From its earliest days, the CPR has had an input from, and been influenced by, Scotland, both in personnel and in the names given to places along the line. As someone who lives in Scotland, I am merely following in the footsteps of those who helped to shape the railway!

If you mention the Rockies and Canadian Pacific, virtually everyone thinks about the line through Banff and Kicking Horse Pass, but there is another line in Southern Alberta that runs through the Rockies. Known by railfans and the people who work on it as 'The Crow', this is the line that runs west from Lethbridge and crosses the Continental Divide in Crowsnest Pass on its way to Fort Steele. Compared with the main transcontinental line further north, the Crowsnest line is not nearly as busy but it runs through some equally photogenic scenery and one thing in its favour is that it is not situated in an area that has a high number of visitors. It is a fascinating line and well worth a visit.

I could not have even contemplated producing this book without all the help that I have been given by so many. I would never have been in the right place at the right time. Telling certain people in the Rockies that I had travelled from Scotland has been like being given a passport to go and do things that are out of bounds to the majority of visitors, even to Canadian railfans, and I am truly grateful for all the assistance that has been offered over the years. My special thanks go to Mike and Anni

Introduction

Auffray from whom I learned so much about Crowsnest Pass and especially to Josh Soles for keeping me so well informed as to what trains were running; without his input I would not have been nearly so successful. I am very grateful to Key Publishing for giving me this opportunity to record some of my photos from my times in Alberta and British Columbia, and lastly for the great company on earlier visits of my son Chris and daughter Laura. Living with a railway enthusiast husband requires a very special person so I have to thank my wife Irene for all her invaluable help and support over the years as without her assistance, this book would never have been achieved.

Ian Lothian
Larbert

Canadian Pacific 8607 and Kansas City Southern 4831 depart from Banff on 21 September 2015 with a westbound mixed freight.

Chapter 1
Exshaw to Canmore

The Canadian Pacific Railway was the means that united all the provinces of Canada; British Columbia was persuaded to join the Canadian Confederation in 1871 with the promise that a railway would be built that would link Canada from coast to coast. The railway was originally planned to run further north and go through Yellowhead Pass but there was a problem for the politicians. In the USA, the Great Northern was building its line westwards across Montana and was proposing a series of branch lines that would serve southern Alberta.

The Canadian politicians of the day thought that there was a possibility that Southern Alberta would vote to become part of Montana and so the railway was built much further south, passing through Calgary as it struck out towards the Continental Divide, which was a lot more difficult to operate because of its route through the mountains. Calgary is just under 3,500ft above sea level and over the next 50 miles to Exshaw over the High Prairies, it gains a further 804ft in elevation. Canadian Pacific divides its lines into sub-divisions and the line from Calgary to Field at the foot of Kicking Horse Pass is called the Laggan Sub. Exshaw is where the High Prairies meet the mountains and where this journey along the Canadian Pacific Railway to Golden British Columbia will start. Canadians have different names for various railway subjects and I will use the Canadian terms along with an explanation throughout this book.

Exshaw has a small yard and a pair of locomotives are outbased each week at Exshaw to work the local trip freights to the three adjacent manufacturing plants. There are two Baymag plants, one each side of Exshaw, which combine to produce magnesium oxide, a substance that is widely used in a variety of medical products. At Exshaw, close to the railway is a large Lafarge cement works, which was originally built in 1906. The plant has been modernised and expanded over the intervening years and is currently the largest cement plant in Canada. The site it has at Exshaw was chosen because of the availability of a vast supply of high quality limestone, power from a hydroelectric plant near Seebe (just to the east) and most importantly, because it was beside the Canadian Pacific Railway's main line, making it easy to transport its variety of finished products. The works is connected to the main network by a spur line from the east end of Exshaw Yard. Just to the east of Exshaw is a passing place on the otherwise single-track main line. In the UK, passing places on single-track lines are called loops; in Canada they are called sidings, which is the term that will be used in this book. The disastrous flooding caused by the severe weather in the summer of 2013 resulted in Lafarge supplying many thousands of tons of stone that was used to create a flood defence for the housing, the works and to ensure that neither the highway nor the railway would again be flooded if a similar period of intense rainfall was to reoccur.

The railway runs past the works and yard at Exshaw and curves sharply to run westwards alongside the south side of the Bow River and it follows the Bow River Valley to Lake Louise Village. As the line runs west, it starts to cross Gap Lake on a causeway and at the eastern end there is the junction for the spur line to the other Baymag plant. At the west end of Gap Lake, the line threads its way through a narrow gap between the mountains and that is where the next location on the railway got its name. Gap has a passing siding and a couple of other sidings where, on my visit in 2013, stone was being loaded and taken by rail to shore up the river bank at Massive, to the west of Banff, which had suffered storm damage. The stone was also used at the passing siding at Massive, which was having its length extended so that longer trains would be able to cross there.

In 2013, there was a period of extreme weather and the amount of water that fell on the mountains brought huge quantities of rocks down and caused a vast amount of damage to the railway and the local roads. The stone that was brought down at Gap was the source of the stone used to repair and strengthen the side of the line where the flooded Bow River had eroded the banks and caused damage to the railway's permanent way. I was able to record a stone train being loaded at Gap and several of the workings between Gap and Massive while I was there later in 2013.

The railway line from Exshaw to Canmore, as well as running beside the Bow River, also parallels Highway 1A for most of the way, the exception being at Gap Lake where the railway crosses over to the south side while the road runs alongside the lake's northern shore. Photography from Exshaw at a multitude of locations is possible from beside the highway for the majority of the distance until the outskirts of Canmore are reached, after that there are still some good locations but access to the line is more restricted.

A pair of Canadian Pacific Railway's GP38-2 locomotives were outbased at Exshaw to serve the local traffic from the Lafarge Works at Exshaw and the two local Baymag plants. On 17 October 2007, 3085 in an older livery and 3044 in the then latest red livery wait in Exshaw Yard with a rake of cars that will soon be taken to the adjacent Lafarge works. The GP38-2s are a four-axle 2,000hp locomotive used on local freight services on the CP network. Canadian Pacific purchased 115 of them from General Motors' Electro-Motive Division between 1983 and 1986.

On the single-track main line, CP 8948 and 9723 pass Exshaw with an eastbound manifest to Alyth Yard in Calgary on 19 October 2012. 8948 is a six-axle 4,400hp locomotive built by General Electric in 2012 in its 'Evolution' series, while 9723 is an older 4,400hp AC4400CW built by GE in 2003.

On 11 October 2011, a different pair of the 2,000hp GP38-2s were working at Exshaw. 3134 and 3030 have collected a rake of empty cars that they will shortly be taking to the Lafarge works to be loaded. 3134 was the penultimate GP38-2 to be constructed.

After depositing the empty cars in the Lafarge works, 3134 and 3030 are returning to Exshaw Yard with loaded cars that will later be collected and taken to Calgary by an eastbound manifest that was due to arrive at Exshaw in the late afternoon.

A westbound mixed freight, a 'manifest', approaches Exshaw on 15 October 2012 behind a pair of AC4400CWs; 9615 was built in 1997 for CP while the second locomotive, blue 1055, which CP took on a long lease, was built in 2004 to CP's specification for CEFX (a leasing company). They are passing GP38-2s 3034 and 3046, which were making up a train in the yard at Exshaw.

A noticeable change to the locomotive scene at Exshaw was the replacement of the pair of GP38-2s by two of the new General Motors GP22C-ECO locomotives. The donor locomotive bodyshells were old withdrawn GP40s and GP40-2s that were rebuilt with new engines and electrics as the main features. On 21 September 2015, 2240 and 2316 are seen stabled in the yard at Exshaw.

One of CP's few remaining SD 40-2s, 6045 is waiting in Exshaw Yard on 30 May 2017 with a snowplough and spreader which it had been taking from Lake Louise to Calgary for maintenance on 28 May. It had developed a fault and ended up in the yard at Exshaw until fitters came out from Calgary to carry out the necessary repairs. The SD40-2 was a class of six-axle 3,000hp locomotive that was built between 1972 and 1989; 4,175 in total were built for American railroad and Canadian railway companies. CP purchased a total of 486 over the years but by 2017, only a handful still remained in active service.

A view of part of the Lafarge Cement works at Exshaw. The plant has its own locomotive, 1749, which is a GM Class GP10 and it is seen near the front of the plant on 19 October 2012. 1749 is finished in a rather attractive black and green livery and was photographed from the adjacent Highway 1A.

On 31 May 2017, the two Exshaw based GP22C-ECO locomotives were 2233 and 2249 and they were caught by my camera as they exited the Lafarge works with a train of loaded hopper cars and started to cross Highway 1A. SD40-2 6045 can be seen in Exshaw Yard with the snowplough and spreader.

The weather in the Rockies can vary considerably from day to day and it is quite possible to have a mix of all four seasons in one 24-hour period. On 21 October 2012, GP38-2s 3034 and 3046 are seen at Exshaw after an overnight fall of snow and with the temperature a rather cold −17°C at 10.00am.

CP 8731, a 2005-built ES44AC, leads an eastbound mix of empty tanks and double stack containers at Exshaw on 30 May 2017.

A pair of ES44ACs, 8757 and 8811 run alongside an unusually still Bow River as they approach Exshaw on 15 October 2007 with an eastbound intermodal going to Calgary.

At the beginning of July 2010, Rocky Mountaineer Railtours were experiencing a motive power shortage due to several locomotives within its fleet having developed problems. The solution was to hire in a CP SD40-2 and early in the morning of 4 July 2010, the empty stock from Calgary to Banff after its overnight servicing is seen shortly after passing Exshaw as it runs westwards alongside the Bow River with RM 8012, a rebuilt former Canadian National GP40, leading CP 5960.

An eastbound train of empty grain cars is approaching Exshaw on 15 October 2007 with 1998-built AC4400CW 8559 leading GE 2006-built 4,400hp ES44AC 8788.

Canadian Pacific in the Rockies

A westbound loaded oil train passes Exshaw on 30 May 2017 with CP 9654 and 8610 leading and 9702 as distributed power on the rear.

An early start on 15 October 2007 enabled me to catch this Vancouver to Toronto intermodal train just after the sun had cleared the mountains to the east as it ran alongside the Bow River to the west of Exshaw with three recently built ES44ACs as power, 8757 leading 8811 and 8802.

CP 2249 and CP 2233 bring loaded freight cars down the spur line that runs from Gap Lake to the Baymag plant, which produces high-grade caustic calcined magnesia, that is situated to the west of Exshaw on 30 May 2017.

CP 2249 and CP 2233 cross Highway 1A between Exshaw and Gap with loaded freight cars from the Baymag plant on 30 May 2017. Note the lack of barriers; the crossing is only protected by flashing lights and relies on the crew members to halt any road vehicles.

On 19 October 2012, GP38-2s 3046 and 3034 wait to join the main line at Gap Lake with a couple of loaded hoppers from Baymag that they will shortly propel to Exshaw.

On 11 October 2011, CP 3030 and 3134 have run onto the main line at Gap Lake with loaded railcars from the Baymag plant. They will now slowly propel their train back to the yard at Exshaw.

CP GP38-2s 3134 and 3030 are halfway between Gap Lake and Exshaw as they propel their rake of loaded railcars with a shunter riding on the front of the leading car in touch with the locomotive crew by radio on 11 October 2011.

SD 40-2 5743 and GP38-2 3070 were in a siding at Gap on 26 September 2013 with a rake of old side tipping stone carriers that have been loaded and will soon be taken along the line to Massive, west of Banff, where the passing siding was being lengthened and the stone was being used to strengthen the ground along the bank of the Bow River.

8738 and 9612 wait in the siding at Gap on 4 July 2010 with a westbound loaded grain train as an eastbound double-stack intermodal passes. They will be able to proceed once the intermodal has cleared the single line from Canmore.

CP 8770 leads an eastbound empty potash train at Gap on 4 July 2010 with the impressive Rundle range of the Rockies in the distance.

8859 and 9555 pass Gap on 28 September 2013 with grain empties on their way back to the prairies for reloading.

Chapter 2
Canmore to Banff

After leaving Calgary and its suburbs behind, once the railway enters The Rockies, Canmore is the largest settlement that it will encounter for more than 200 miles. The unfortunate thing is that although Canmore has local industry, there is no traffic either delivered or uplifted on a regular basis by rail. Canmore has an engineer's siding and a long passing siding and that nowadays is the only involvement that the Canadian Pacific Railway has with the town. In Canada, as in the USA, there is a requirement for a locomotive to sound its horn when approaching road crossings in the format of two long, one short and one long. As Canadian Pacific Railway's trains run 24 hours a day, seven days a week, in order to lessen the sound of trains in the middle of the night, Canmore and the CPR combined to make their road crossings as safe as they could be so that the trains did not have to sound their horns when approaching crossings. This has involved many varied things from building a built-up central reservation on roads to stop any driver from zig-zagging round the barriers after they have dropped, to installing traffic signals that send a red light to road travel. By making the crossings safer, trains no longer sound their horns, which must have greatly improved conditions for those who live adjacent to the crossings.

Between Canmore and the approach to Banff, the railway cannot be reached from the Trans-Canada Highway and photography on this part of the line is not possible. Just to the west of Canmore, both road and rail enter Banff National Park and all visitors using the roads must be in possession of a Parcs Canada Pass, which has to be visibly displayed in your vehicle. For westbound trains, it is a steady climb up from Canmore with the line gaining another 100ft in height over the 17 miles, a total of 270ft since passing Exshaw. As the line is following the Bow Valley, some parts of the line have gentle gradients but with several parts having some short but steeper inclines which, when coupled with some of the necessary curvature, does not make for fast running.

Banff has a long passing siding, two other sidings and possesses a 'wye' (a triangle), which is used to turn both locomotives and passenger trains that are short in length. Banff has a road crossing at each end of its passing siding but unlike the crossings at Canmore, at Banff the locomotives still sound their horns to warn road traffic. The Rocky Mountaineer tourist service uses the depot at Banff, starting and ending its two day trips to and from Vancouver at Banff with the train then running as empty stock in both directions between Banff and Calgary for overnight servicing of both the passenger accommodation and the locomotives. The Royal Canadian Pacific often runs out to Banff in a morning and returns to Calgary in either the late afternoon or in the evening when it does its day trips to Banff. Its passengers are usually taken to the Banff Springs Hotel for a meal and if the weather permits, a round of golf on the mountain course. Banff Depot was declared a Heritage Station by the Government of Alberta in 1991.

Banff has some excellent locations for photography and there are many different mountain backdrops to choose from. The photograph of the westbound intermodal that was taken from the summit of Sulphur Mountain was made possible by going up in the gondola. If one uses a good telephoto lens, it is possible to shoot a westbound over the shoulder of Tunnel Mountain and in the morning, it is also possible to photograph an eastbound as it runs down the Bow Valley and approaches Banff. The town of Banff is a resort town with numerous places to stay and/or eat. The town was given its name by George Stephen, the CPR's first chairman who named it after his birthplace in Scotland. Banff National Park is a United Nations World Heritage Site and well worth a visit and if you want to see plenty of wildlife, just follow the railway!

If anyone looks at some of my photographs and thinks that getting to the various locations involves a lot of walking or climbing, this is simply not the case. The vast majority of photographs in this book were taken only a few yards from my hire car and the one thing that you do not do over there is leave your car and walk any distance unless you are in a group; the National Parks have a lot of wildlife that can seriously injure and even kill so keeping a good lookout at all times is advised.

Above: Canadian Pacific's AC4400CW 9738 and ES44AC 8781 lead a Vancouver to Toronto priority intermodal away from Canmore towards Gap on their way to Calgary on 31 May 2017.

Opposite above: With a magnificent mountain backdrop, a pair of CP locomotives, 9838 and 9728, leave Canmore as they head eastwards towards Gap with a train of empty grain cars on their way back for loading somewhere on the prairies.

Opposite below: A train of empty ethanol tank cars leaves Canmore for Calgary behind three CP locomotives, 9524 leading 9506 and 9542 on 14 June 2009. The first two have the old 'CP Rail' and the twin flags logo on the bodysides while the third locomotive has the wording 'Canadian Pacific' and wears the Beaver logo, known affectionately by railfans as 'The Golden Rodent'.

Canadian Pacific runs a luxury charter train called the Royal Canadian Pacific, this using dedicated locomotives and a train of coaches that have been rebuilt and fitted out internally to the highest standard, making it a five-star hotel on wheels. Canadian Pacific staff call the coaches 'The Business Cars'. On 10 October 2010, the train is seen as it leaves Canmore with FP9 4107 in its heritage paint scheme leading GP38-2 3085 in the same livery. The GP38-2 was paired with the FP9 to help with braking on steep gradients as well as providing power to help haul the train. 4107 was built in 1957 and is a 1,750hp machine.

CP SD40-2 5796 leads Rocky Mountaineer's 8011 at Canmore on 14 June 2009. The Mountaineer had been serviced overnight at Calgary and then runs to Canmore where it picks up its staff before running on to Banff where its passengers will board for their two-day journey to Vancouver. The mountains seen over, and to the right of, 5796's cab are the Three Sisters; the three peaks associated with the town of Canmore.

Late in the evening on 20 May 2017, 8621 leads a westbound intermodal at Canmore with CP 9823 working as distributed power halfway along the train.

During many of my early visits to the Laggan Subdivision, the only locomotives seen on CP freights were either owned by CP or were CP leased locomotives. On 16 June 2009, it came as a surprise to see this very short in length freight going westwards through Canmore with CP 8530 leading Canadian National's 5600, a 4,000hp SD70I built in July 1995.

Late September 2018 saw a prolonged spell of poor weather with unusually low temperatures, low cloud and some significant snow. 27 September had been a day with low cloud and rather poor light but for a short while in the late afternoon, it brightened up with around an hour's sunshine. Right in the middle of this period of better light, CP 9700 and 8858 climb up through Canmore at the head of a loaded grain train on its way from Calgary to Vancouver.

Blue leased 1050 leads CP 9713 at Canmore on an early evening empty grain train going to Calgary before being taken further east to the prairies to be loaded with more grain for export.

This was a very unusual movement, which I was fortunately able to photograph. SD40-2 6045 had run from Calgary to Lake Louise to collect the spreader and snowplough that are based there and are used to clear the line over Kicking Horse Pass in winter. 6045 was supposed to take the snow clearing equipment to Calgary for its annual maintenance, but on its return journey it developed a fault, which is how it came to spend several days in the yard at Exshaw waiting to be repaired. 6045 was photographed as it approached the first level crossing in Canmore on 28 May 2017

A closer look at the plough and the spreader behind 6045 at Canmore on 28 May 2017.

CP 8749 and CEFX 1041 were running downgrade as they passed Canmore on 23 May 2017 with a Vancouver to Calgary intermodal.

On the morning of 23 May 2017, Canadian Pacific ran the Royal Canadian Pacific from Calgary (where it is normally based) to Banff on a day tour. In the evening, the RCP is seen as it approaches Canmore returning to Calgary being hauled by the three rebuilt FP9s with 4107 leading 4106 and 1401.

An eastbound empty grain train unusually hauled by three American BNSF locomotives has stopped on the single-track main line as a westbound loaded potash slowly runs into the passing siding at Canmore with CP 8841 and 8915 as the two locomotives on the head end on 30 May 2017. The potash train, 170 loaded hopper cars, had two further locomotives, 9813 in the middle of the train and UP 5341 on the rear, all controlled by the engineer on the front end in 8841.

When I first saw this train approaching in the distance I thought I must be seeing things as it was such an unusual sight. Three BNSF locomotives, 1086, 5203 and 4865, approach Canmore on 30 May 2017 with an empty grain train from Vancouver that was going to Lloydminster for loading with grain for export with the three BNSF locomotives working right through to Lloydminster.

After an overnight fall of snow, 8533 and 8537 are working an empty grain train from Vancouver to Calgary at Canmore on 20 October 2012.

An overnight fall of snow on the mountains helped to provide a stunning backdrop as 9813 and 8844 approached Banff East Siding at the head of a westbound potash with 170 loaded hoppers and a further two helper locomotives, one in the middle and the other on the rear on 22 May 2017.

On a beautiful Fall morning, which helps to show the scale of the surrounding mountains, 9830 and 8627 approach Banff East Siding with a westbound intermodal from Calgary to Vancouver on 17 October 2012.

It is almost an impossibility to be able to capture the full length of a freight train in a photograph and this perhaps shows the challenge facing any photographer trying to achieve this. A lengthy double-stack intermodal has two locomotives on the front, 9813 and 8844, as it approaches Banff East Siding while the rear is out of sight in the distance. The road is Highway 1, the Trans-Canada Highway, and the photograph was taken from near the summit of Sulphur Mountain, Banff.

On 29 June 2005, Canadian Pacific's Hudson 2816 was working a four-day excursion from Calgary to Vancouver; the first day taking the train as far as Lake Louise village. I photographed it first as 2816 approached Banff; the sound of its hooter echoing off the mountains was something that I will always remember. CP 2816 is also known as the 'Empress' and is a Class H1B 4-6-4 that was built by the Montreal Locomotive Works in 1930. After a major restoration, during which she was converted to being oil-fired, she worked excursion traffic from 2001 until 2012 when CP terminated the steam programme, although 2816 was steamed again in 2020 but is currently in store in Calgary.

A pair of 4,400hp AC4400CWs, CP 9615 and CEFX 1055, haul a westbound mixed freight upgrade towards Banff on 15 October 2012.

Two of Canadian Pacific's FP9s, 4106 and 4107 bring a Royal Canadian Pacific Calgary to Banff day excursion upgrade from Canmore towards Banff on 27 September 2013.

This was the only occasion that I managed to photograph the Royal Canadian Pacific when GP38-2 3084 was leading. I had just arrived at Banff East when the RCP came into view and stopped to wait for a westbound to pass before it could continue its journey to Calgary. 3084 was leading FP9 1401 on 28 June 2005.

On 29 June 2005, an eastbound mixed freight departs from the siding at Banff East and resumes its journey to Calgary. The leading locomotive, CP 9664, is another AC4400CW while the second locomotive, 9128, is a General Motors SD9043MAC, a 4,300hp machine that was built in the mid-1990s but the class proved to be unreliable and they all ended up in long-term storage after a working life of only around ten years. In the past two years, a programme to convert them into SD70ACu locomotives by Progress Rail, in Indiana, has got under way and the rebuilt locomotives are numbered in the 7000 to 7059 range.

In the late evening of 29 June 2005, the Rocky Mountaineer passes Banff East on its way to Calgary for servicing with RM 2011 leading RM 2012.

Almost 12 years later, I photographed the Rocky Mountaineer at the same location but by 20 May 2017, the locomotives and coaches were all in a completely new livery. 8013 and 8016 pass Banff East on their way to Calgary after depositing their passengers at the depot at Banff. At just over 8,000ft, Sulphur Mountain is the backdrop.

The GM SD9043MACs might have been an unsuccessful design but they did look very impressive locomotives. 9113 and 9111 pass Banff East on 29 June 2005 with grain empties from Vancouver on their way to Calgary.

Canadian Pacific's Hudson 2816 is again seen in action, this time on 1 August 2006 as it departs from Banff with an excursion returning to Calgary.

I photographed the Royal Canadian Pacific at Banff on 23 May 2017 when it had all three of CP's FP9 locomotives hauling the train; the 9,616ft high Mount Bourgeau forms an impressive backdrop.

Hudson 2816 waits at the depot at Banff on 1 August 2006 prior to departing with an excursion to Calgary; a couple of old boxcars with old-style wording help to add to a timeless scene.

The Royal Canadian Pacific had just arrived at Banff from Calgary on 27 September 2013 and its locomotives would shortly stable the business cars in a siding and then run round after which the train would wait for its evening departure time to return to Calgary. The two mountains in the distance are the 9,724ft Mount Inglismaldie on the left and Mount Girouard on the right, which at 9,826ft is the highest peak in the Fairholme Range. Mount Inglismaldie was named after Inglismaldie Castle near Laurencekirk in Scotland.

The Royal Canadian Pacific has a range of beautifully restored and rebuilt passenger cars, all of which are named. This is *Van Horne*, CP 77, named after the first general manager of the Canadian Pacific Railway. It was built in 1927 at Montreal, rides on six-axle bogies and features both a lounge and dining area. This photograph was taken at Banff on 23 May 2017.

Business car CP 74 is named *Mount Stephen* after the Canadian Pacific Railway's first president Sir George Stephen, 1st Baron Mount Stephen. It is the tail end lounge car with an open balcony so that its guests can, weather permitting, fully enjoy the splendid scenery and I photographed it at Banff on 23 May 2017.

Another pair of the short-lived SD9043MACs, 9103 and 9130 pass the depot at Banff with a westbound intermodal on 26 June 2005.

CP 8643 and two other AC4400CWs pass Banff with a Calgary to Vancouver intermodal on 28 June 2005.

Canadian Pacific started to rebuild and upgrade its earlier built AC4400CWs in a programme that started in 2017. On 27 September 2018, 8125 was leading CEFX 1031 as the pair arrived at Banff on a day with low cloud and snow flurries on an eastbound train of empty oil tanks returning to Calgary. When the locomotives were rebuilt, they were given new numbers, 8125 is the former 9641 and the rebuilt locomotives are classed as AC4400CWM.

After serious flood damage from the Bow River at Massive and with CP lengthening the siding there, there was a requirement to strengthen and build up the river banking. CP ran a series of trains that loaded with stone at Gap and then had the contents tipped at Massive with haulage being provided by GP38-2 3070 and SD40-2 5743. This shows the pair passing Banff on 26 September 2013 with their train of loaded old side-tipping wagons.

I was able to photograph several of these stone trains while I was at Banff in 2013 and this shows 5743 and 3070 entering the siding at Banff when they were returning to Gap for loading on 27 September 2013. The opportunity to photograph a train with an SD40-2 leading was too good an opportunity to not take advantage of.

The Rocky Mountaineer waits at Banff prior to departing for Kamloops on one of its last tours of the operating season on 29 September 2013 with RM 8013 on the front.

RM 8013 and 8016 depart from Banff with a Rocky Mountaineer two-day tour to Vancouver on 21 May 2017. In 2017, the departure time from Banff was altered from 09.00 to 08.00, which made photography more difficult as, at that time, the sun was not as far round and had only just cleared the mountains minutes before departure time.

ES44AC 8834 leads an older AC 4400CW westwards through Banff with a priority intermodal to Vancouver on 14 October 2010.

RM 8013 and 8016 are arriving at Banff in the evening on 20 May 2017 at the end of a two-day Rocky Mountaineer tour from Vancouver.

Another view looking west from Banff as CP's AC 4400CW 8514 with SD9043MAC 9101 take an eastbound intermodal into the passing siding at Banff on 27 June 2005 with Mount Bourgeau again prominent on the skyline.

Chapter 3
Banff to Divide

After passing Banff, the railway runs on the south side of the Vermillion Lakes while Highway 1 runs west to the north of the lakes. The railway curves round and passes under the Trans-Canada Highway where the slip road leaves Highway 1 and the Bow Valley Trail (Highway 1A) starts. From this point, the road and the railway are never far apart and there are several places for photography before the road reaches the Castle Mountain Junction. The line is still steadily climbing; in the 35 miles from Banff to Lake Louise Village the line climbs another 600ft.

After Banff, the first crossing siding is at Massive, just to the east of the Castle Mountain road crossing and this siding was lengthened in 2013 when the rock trains were being used to strengthen the bank of the Bow River. After the road crossing, a decent location for a train climbing up from Massive, the line continues westwards and by driving up the Bow Valley Trail, it is not long before the car park at the Storm Mountain lookout is reached. This is a nice location; one of the few where a good photograph of both an eastbound and a westbound can be obtained. It might seem strange to say that when leaving the car to take a photograph, especially if there are only a few other people about, great care has to be taken and frequent looks around are necessary because of the wildlife in the National Park. I have come across bears, bull elks and a moose in this area and one has to remember that more people are killed by moose than by all the other animals put together.

After passing Storm Mountain, the line runs out of sight at a lower elevation than the road until the Baker Creek Mountain Resort is reached, after which the railway is once again close to the road and shortly afterwards there is the passing siding at Eldon. At its east end, there are also other sidings and a wye for turning. As a result of the height of the trees at the lineside, photography without too many shadows getting in the way is a challenge, but it is possible.

As the line proceeds westwards, it is not long before it reaches Morant's Curve, a location where the line curves round alongside the Bow River. The location is named after Nicolas Morant, at one time the publicity photographer for the Canadian Pacific Railway. He used this location to take stunning photographs and it was his opinion that the scenery at this location was as good as any that could be found anywhere in the Rockies. At one time, the vegetation was always cut back here so that visitors could see the trains and photograph them just as Nicolas Morant had done. Unfortunately for the railfans, Parcs Canada has adopted the ridiculous position of forbidding trees to be cut in the National Parks with the result that here and elsewhere, the trees have grown and are stopping visitors from seeing the very views that made them want to come. When the Trans-Canada Highway was converted to a dual carriageway, many thousands of trees were cut yet, at important viewpoints, it will not sanction the removal of just a couple of dozen, an attitude that is puzzling. Photography at Morant's Curve is best in the early morning with an eastbound train.

After Morant's Curve, it is only another couple of miles before Lake Louise Village is reached. The location is the village of Lake Louise and not the lake, which is a distance uphill to the south. At Lake Louise Village, the line ceases to be a single track with two lines all the way up to the Continental Divide at the location that the railway simply calls 'Divide'. The original line climbed initially at a much gentler gradient but the last mile up to the summit was much steeper, which often caused delays when trains got stuck. To improve capacity and to lessen the steep finish of the older line, a new line was constructed and opened in 1981. Leaving the original line at Lake Louise Junction, it climbs all the way to the summit on a consistently lower gradient and this line is now used for virtually all the westbound freights with the original line from Divide to Lake Louise used for eastbound traffic. One exception is the westbound Rocky Mountaineer, which can, on certain tours, pick up passengers at the former Lake Louise station on the original line; with its two diesel locomotives, the last mile or so is not a problem to this train. There are several locations at Lake Louise, on the climb up and at Divide, where photographs can be taken; again it is essential to keep a good lookout for wildlife.

Above: With Sulphur Mountain on the skyline, AC4400CW 9680 leads a loaded potash westwards at Sawback on 2 July 2005.

Left: CEFX 1052 and CP 9677 approach Castle Mountain crossing on 26 September 2015 with a loaded grain train on its way to Vancouver.

CP 9673 leads CEFX 1044 and CP 9549 at Castle Mountain crossing with a westbound intermodal from Calgary to Vancouver on 25 June 2005.

With two brand new ES44ACs, 8928 and 8927, on the front, and sister locomotive 8924 on the rear, a trial to see how the new locomotives would perform on a westbound loaded grain train took place on 9 October 2011. The locomotives had also been specially cleaned in order to obtain publicity photographs and I photographed this working as the train approached Castle Mountain crossing.

On 31 May 2017, Canadian Pacific 8800 and 9357 take a loaded ethanol train westwards as they approach the Storm Mountain lookout. This illustrates how dense the forest can be on either side of the line in places and why it is neither a good nor sensible idea to venture into those areas with the amount of wildlife that exists in Banff National Park.

CP 5796 and RM 8011 run alongside the Bow River as they approach the Storm Mountain lookout with the Rocky Mountaineer on its way from Banff to Kamloops on 14 June 2009.

CP 9638 and 9373 pass the Storm Mountain lookout with a loaded grain train on its way to Vancouver for export on 26 September 2015.

CP 8760 and 9714 are taking an eastbound intermodal from Vancouver to Calgary on 13 October 2007 as they approach the Storm Mountain lookout.

A long westbound manifest is waiting in Eldon Siding for a new crew to arrive from Field late in the evening of 25 June 2005 with CP 9598 leading 9830.

On the lovely summer evening of 24 June 2005, CP 8536 leads a westbound mixed freight alongside the Bow River on the approaches to Morant's Curve; a location named after and made famous by a former CP publicity photographer, Nicolas Morant, who took many photographs at this location which had, in his opinion, views that were as good as any that you could find anywhere in the Canadian Rockies.

On 29 June 2005, Canadian Pacific's Hudson 2816 performed a run past at Morant's Curve and I was fortunate to be able to photograph this as 2816 blackened the sky as it blasted round the curve on its way to its overnight stop at Lake Louise Village.

A track evaluation train hauled by one of Canadian Pacific's then last few remaining GP35 locomotives, 8217, passes Morant's Curve on 29 June 2005 on its way west from Calgary to Golden.

A loaded grain train starts to round Morant's Curve on 27 May 2017 behind CP 8731 and 9677 with 9618 way back as distributed power on the rear.

The spectacular mountain backdrop at Morant's Curve lets any visitor see why Nicolas Morant used this location for publicity photographs. An empty grain train on its way to Calgary runs round the curve alongside the Bow River with CP 9813 leading CEFX 1056 on 17 October 2010.

CP 9516 and 8735 are running alongside the Bow River on a beautifully clear but very cold and frosty morning of 13 October 2007 with a Vancouver to Toronto priority intermodal.

Almost ten years later on 28 May 2017, I was able to photograph this empty potash train at Morant's Curve with ES44AC 8769 providing the power and with one other locomotive in the middle of what was a very long train.

A Loram grinder, used to grind out any small cracks in the railhead, works its way around Morant's Curve on 25 May 2017.

The grain train trial that I had photographed at Castle Mountain crossing is seen again on 9 October 2011 as 8928 and 8927 approach Lake Louise Village on their way west.

Sometimes you just get lucky and are in the right place at the right time! I was at Lake Louise on 5 October 2011 when a people carrier arrived with a relief crew for Train 111, a priority intermodal from Toronto to Vancouver. The train duly arrived about ten minutes later and stopped right in front of me as the crews changed over, giving me the opportunity to take some good photographs of this event.

An eastbound intermodal with CP 8854 leading 8652 splits the signals at Lake Louise on 22 September 2015. This is the point where the new additional track to the summit starts; built to ease the gradient from Lake Louise to the Continental Divide and increase capacity on the line. The new line opened to traffic in 1981 after which the majority of westbound trains use the new line with eastbounds using the older route.

Running on the original line, an eastbound intermodal going to Calgary approaches the junction of the two lines at Lake Louise Village on 13 October 2007 with 9760 leading 9714 and with the impressive mountains that form the Continental Divide on the skyline. The newer track from here to Divide can be seen to the right of 9760.

CP 8516 is leading a heavy westbound grain train on the challenging climb from Lake Louise to Divide on 1 July 2005.

Slightly further up the climb on 14 October 2007, CP 8834 and 9612 are moving their westbound intermodal up towards Divide at a fairly slow but steady speed.

By contrast, RM 8011 and 8016 were climbing to Divide at a good speed with the Rocky Mountaineer on its way from Banff to Kamloops on 14 October 2007. Mount Bosworth, at 9,085ft, provides the mountain backdrop.

The grain trial, using the two new ES44ACs 8928 and 8927 with 8924 on the rear, was climbing to Divide on 9 October 2011 when it was caught in a very fortunate burst of sunshine as it neared the summit on the new line with the older line in the foreground at a much lower elevation at this point. 8927 had been displaying a higher than expected temperature and one of the access doors to the engine compartment had been left open to help with the cooling.

This sulphur train was one that did not quite make it to the top. 8951 was pulling not only the train on 17 October 2012 but also 8566 the second locomotive, which had developed a fault. The train was going slower and slower until with less than a 100 yards to go, it finally stalled and had to wait for additional power to arrive from Field at the foot of Kicking Horse Pass.

This was something well worth waiting for! At just after 8am on 30 June 2005, Hudson 2816 arrives at Divide after climbing up from Lake Louise on the original track. Despite it being only a week from Midsummer Day, at 5,339ft above sea level at that time of the morning, it was very cool but being able to hear and then see 2816 attacking the climb up to the summit is something that I will always remember.

After photographing 2816 at several locations descending Kicking Horse Pass to Field, I was able to return to Divide in time to photograph the westbound Rocky Mountaineer on 30 June 2005 with RM 8011 leading RM 8012.

With 9753 leading 9652, an eastbound empty grain train has completed its climb of Kicking Horse Pass and is about to start the descent from Divide to Lake Louise. Cathedral Mountain at 10,463ft in height is the impressive peak in the distance.

CP 9527 and CP 8520 curve round at Divide and are about to pass under the Trans-Canada Highway (from where this photograph was taken) with another empty grain train heading to Calgary on 26 June 2005.

Chapter 4
Divide to Field

The summit at Divide is 5,339ft above sea level and after the long climb to the summit, westbound trains are now confronted with an even greater challenge, the descent of Kicking Horse Pass. The passing siding at the summit is called Divide at its east end and Stephen at its west end. At Stephen, the line starts to descend past Sink Lake, it crosses the Lake O'Hara road and there is then a short level section when the line runs along the south edge of Wapta Lake, the source of the Kicking Horse River. Highway 1 runs alongside the north shore of the lake and it is possible to photograph up to the west end, after that the railway and the road go their own ways.

When the line was being constructed, it originally took a direct line towards the foot of the 'Big Hill', as that part of the line is known, and it descended at a ruling grade of 1 in 22, slightly more than the recommended maximum grade for the then steam traction. Runaways were common and there were three sidings on the hill that were manned; if a train was running away, it was diverted up one of these sidings to minimise damage to lives and equipment. There were also several operating constraints, the maximum speed allowed for a passenger train was 8mph, and 6mph for freights. The maximum number of freight cars that one locomotive was allowed was five and with five locomotives, the maximum number of wagons in a train was 25.

As traffic grew, this was simply inadequate so something had to be done. Building the line had very nearly bankrupted Canadian Pacific so time was needed to recover and plan what could be done. The idea of running around the mountain sides was abandoned because of the constant risk of avalanches. CP's engineers were sold the idea of a new line down the pass using spiral tunnels as had been built in Switzerland on the St Gottard Railway. Construction of the new line was started in 1907 and completed in 1909 and this resulted in the maximum grade being eased to 1 in 48, meaning longer and heavier trains could now be safely worked over this section of the line.

The line descends from Wapta Lake and soon enters the Upper Spiral Tunnel in Cathedral Mountain. When a train emerges, it is now at a level that is almost 50ft lower and it keeps on descending through Yoho, curves round to the other side of the valley and enters the Lower Spiral Tunnel in Mount Ogden. Again, when it emerges it has lost almost another 50ft in height and it then recrosses the valley to run down to Field at the foot of the incline on the south side of the Kicking Horse Valley. A slide known as 'The Big Slide' resulted in a reinforced avalanche shelter having to be constructed and after passing through this, a westbound train will soon after arrive at Field and this is where the Laggan Sub ends and the Mountain Subdivision starts. Despite modern motive power, there continue to be accidents on Kicking Horse Pass with the last serious one being as recently as 8 February 2019, when a CP loaded grain train ran away and derailed not long after exiting the Upper Spiral Tunnel, unfortunately killing all three of the crew on the leading locomotive.

Field, at the foot of the Pass, is a crew change point and the start of the Mountain Subdivision, which runs as far west as Revelstoke on the west side of Rogers Pass, Canadian Pacific's crossing of the Selkirk Mountains. In steam days, there was a roundhouse at Field where locomotives were based to be used as bankers to help push trains up Kicking Horse Pass. In the early diesel days, there were still banking locomotives based at Field but nowadays, with high horsepower diesels and the ability to have distributed power, the need for banking locomotives to be based at Field is no longer required. Distributed power is when additional locomotives to those on the front of a train are inserted into the middle of a train and/or on the rear and they are all controlled by the engineer (driver) in the leading locomotive by radio signals.

Left: A loaded grain train waits at Stephen, at the west end of the summit siding, on 28 September 2013 with CP 9603 leading two further locomotives.

Below: After waiting for an eastbound to have completed the climb up from Field, 9603 finally has got the road and is starting the slow and cautious descent of Kicking Horse Pass, known to the CP crews as the 'Big Hill'.

With the mountains that form the Continental Divide as a skyline, CP 8834 with 9612 start the descent of Kicking Horse Pass on 14 October 2007 with an intermodal to Vancouver.

On 5 October 2011, the last westbound Rocky Mountaineer of the 2011 season from Banff to Kamloops and Vancouver passes Stephen and is about to start the descent of Kicking Horse Pass.

With their dynamic brakes whining, CP 8928 and 8927 bring the grain train trial slowly down Kicking Horse Pass as they approach the Lake O'Hara road crossing on 9 October 2011.

CP 8723 and 8541 approach the Lake O'Hara road crossing on 26 May 2017 with a train of loaded tank cars for Vancouver.

The steam excursion with Hudson 2816 was to run from Lake Louise to Revelstoke on 30 June 2005. The crew of 2816 had implied that they did not have a vast amount of confidence in the brakes on some of the passenger cars and so were taking things very easily as they approached the Lake O'Hara road crossing.

An empty grain train on its way to Calgary is climbing Kicking Horse Pass on 26 June 2005 with CP 9527 and 8520 having to work hard on the steep climb up from Field.

It is not always a case of sunshine and blue skies in the Rockies. On 25 September 2015, as I photographed 9647 climbing the pass leading an empty grain train, the mist suddenly started to thicken and it started to rain; not at all the conditions that both myself and the crew of 9647 wanted!

8930 is at the head of empty tanks on their way back to Alyth yard at Calgary from Vancouver. The train is running on the south side of Wapta Lake while the Trans-Canada Highway is on the north side. Wapta Lake is the source of the Kicking Horse River and the mountain is Cathedral Mountain, 10,463ft above sea level.

After passing through the Upper Spiral Tunnel, 2816 and its excursion train drift slowly downgrade on the approach to Yoho with part of Cathedral Mountain looming in the background.

A decision by Parks Canada not to cut trees in the National Parks has meant that many famous views can no longer be seen or enjoyed. I photographed this grain train at the Lower Spiral Tunnel on 26 June 2005 but unfortunately trees have since grown and it is no longer possible to see a train using the Lower Spiral Tunnel. Exiting the tunnel as its train was running overhead, CEFX 1054 was leading CP 9607 and 9117 with a loaded grain train going to Vancouver.

Having descended from the Upper Spiral and used the Lower Spiral in Mount Ogden, 2816 and its excursion train have just passed under the Trans-Canada Highway and are approaching Cathedral as they continue to make their way down Kicking Horse Pass on 30 June 2005.

A problem that faced Canadian Pacific for many years was what is known as the 'Big Slide'; an area 2.9 miles east of Field where avalanches were a constant threat, causing derailments and damage over the years after the line was opened. The eventual solution was to construct a reinforced avalanche shelter that was completed in 1988. On 1 July 2005, the Rocky Mountaineer, hauled by RM 8013 and 8014, is about to enter the shelter as it makes its way up Kicking Horse Pass on its way to Banff.

The crew of 2816 had worried about an apparent lack of brakes so had stopped on Kicking Horse Pass and had screwed down various handbrakes; this was effective in reducing the train's speed but did have the result that when the train came off the 'Big Hill' and approached Field, some brakes were really smoking and a long stop had to be made at Field to repair the damaged brakes.

The grain train that I had photographed as it exited the Lower Spiral Tunnel is seen as it arrives at Field on 26 June 2005 with CEFX 1054 leading CP 9607 and 9117.

A double stack intermodal on its way to Vancouver arrives at Field on 13 October 2007 behind CP 9736 and 8733.

The Rocky Mountaineer from Banff arrives at Field on 28 May 2017 with RM 8012 and 8019 providing the power.

The sun had broken through the morning mist that was hanging over the Kicking Horse River in time for me to photograph CEFX 1042 and CP 9582 arriving at Field in sunshine with a 170 wagon loaded potash bound for export from Vancouver on 8 October 2011.

Another westbound loaded potash has safely negotiated the descent of Kicking Horse Pass and has arrived at Field on 13 October 2007 where a crew change will take place before it continues its journey. The 170 loaded cars had CP 8522 and 9608 on the front, 8751 and 9587 in the middle and one other on the rear.

Field is surrounded by some very photogenic mountains; this shows the old depot with CP 9678 stabled in a siding on 13 October 2007 waiting for its next duty. Towering over Field is Mount Stephen, which is 10,495ft in height and named after George Stephen, the first president of the Canadian Pacific Railway.

After a crew change, CEFX 1042 and CP 9582 start to move their potash train westwards and pass another blue and red pairing, CEFX 1056 and CP 9628, which were waiting for the potash train to clear the single line before they were able to start their climb of Kicking Horse Pass on 8 October 2011 with an empty grain train.

One of the things that happen on railways in Canada is that when trains cross with one in a siding on single-track lines, the conductor of the waiting train gets out and does a visual check of the passing train. This is illustrated in this photograph, which shows CP 8723 and 9815 arriving at Field while the conductor from 8749, which is leading 1041, prepares to do her check on the passing train before her train starts its ascent of Kicking Horse Pass.

A view of a busy Field yard on 4 August 2006. In steam days, the Field roundhouse was situated in the foreground and it is still possible to make out where the shed roads were situated. The mountains in the distance are the Van Horne range, named after the Canadian Pacific Railway's first general manager.

In 1990, Canadian Pacific abolished the use of a caboose on the rear of freight trains so it was a pleasure to see 9655 with a white caboose at Field on 26 May 2015. Caboose CPA 20985 is used to transport staff working on the track to areas where there is no road access and on that day, it was about to go and collect staff who had been working on a remote area of track on Kicking Horse Pass.

Chapter 5
Field to Golden

The line from Field to Golden is the start of the Mountain Subdivision and the line now follows the Kicking Horse River all the way to Golden. The real difficulties that face Canadian Pacific's operating department can be easily understood by looking at the distances and differences in height on both sides of the Rockies. The distance from Calgary to the summit of Kicking Horse Pass is approximately 140 miles with a difference in height of 1,913ft. On the western side, Golden to the summit of Kicking Horse Pass is 58 miles and the difference in height is 2,739ft. The Kicking Horse River is one of the fastest flowing rivers in Canada and between Wapta Lake and Golden there are three major waterfalls as well as many rapids, with the result that the line's descent from Field to Golden has several lengths with steeper gradients.

Building the line very nearly bankrupted the CPR and as with the line down Kicking Horse Pass, the line on towards Golden had to be constructed as cheaply as possible and as fast as possible in order to run trains that would generate income. The original alignment between Field and Ottertail was via Muskeg Summit, which involved a steep climb at 1 in 48 for trains going west from Field, and a slightly easier climb at 1 in 55 for eastbounds, but both climbs required the use of pusher locomotives. At Ottertail, the line had to cross the Ottertail River and Valley, which it did on a rather spectacular wooden trestle that was 705ft in length and 122ft high. As the use of iron and steel was costly, and had to be transported over a long distance to this part of the line, the solution was to build bridges of wood using the abundant supplies of timber that existed along the sides of the tracks.

As revenue started to flow in and the financial position improved, the wooden bridges and trestles were gradually replaced and a new line was built in 1902 between Field and Ottertail that is still in use today. The new line increased the distance by just over a mile but it also meant that a much shorter bridge was required at Ottertail with that bridge lasting until it was rebuilt following scour damage in 2014, and it also eliminated the need for pusher locomotives to Muskeg Summit. From Ottertail, the line follows the Kicking Horse River past Leanchoil and then at Palliser, it starts a steep descent to the river bank and winds alongside the river through the narrow Kicking Horse Canyon past Glenogle until it emerges at Golden.

Once past Palliser, there are very few photographic opportunities until the line reaches Golden. Golden is an important location on Canadian Pacific Railway's transcontinental main line, having a small yard immediately after the line emerges from the Kicking Horse Canyon and then a triangular junction with lines leaving the main line, which join and cross the Kicking Horse River before arriving at a large yard and a repair workshop. This line is the Windermere Subdivision that runs south to Fort Steele where the line diverges west to Cranbrook and east to Sparwood. Empty coal trains from Roberts Bank, Vancouver, use this line to return to Sparwood and the opencast mines in the Elk Valley while trains going to Lethbridge travel over Crowsnest Pass. The population of Golden is just under 4,000 but it does have good amenities for visitors. Locomotives and crews are based at Golden and it is still in many ways a railway town.

A Calgary to Vancouver intermodal passes Ottertail on 28 September 2013 with CP 9624 leading 9717 and 8513. The bridge carries the line over the Ottertail River.

The Rocky Mountaineer with RM 8012 leading, then a locomotive that was hired in to help with a power shortage from a leasing company, 2020, and finally RM 8019 pass Ottertail with a Banff to Kamloops tour on 5 October 2011.

CEFX 1042 and CP 9582 pass Ottertail on 8 October 2011 with a heavily loaded potash at a slow speed because of the combination of sharp curves and a steep gradient.

The bridge at Ottertail was damaged by scouring of the former stone piers, seen in the three previous photographs, and new concrete piers were installed. The rebuilt bridge is seen in this photograph as a pair of AC4400CWs, 8531 and 8541, pass Ottertail on 21 May 2017 with a train of loaded tank cars.

A double stack intermodal crosses the bridge at Ottertail with CP 8838 and 9647 taking the train west towards Golden on 23 September 2015.

By 2017, it was quite common to encounter locomotives belonging to Union Pacific but when they did appear it was usually with a CP locomotive on the front. On 23 September 2015, I was fortunate to be able to photograph a mixed freight going west at Ottertail with UP 5545 leading CP 9761.

On 13 October 2007, CP 9736 and 8733 pass Leanchoil with a Calgary to Vancouver intermodal.

The same pairing are seen shortly afterwards as they descend the steep gradient towards the Kicking Horse Canyon near Palliser on 13 October 2007.

A loaded grain train on its way to Vancouver emerges from the Kicking Horse Canyon at Golden on 5 August 2006 with CP 8559 leading 9656 and 9554.

Looking west over the Kicking Horse River at the east end of Golden on 26 May 2017, this photograph shows two trains with CP 9861 and 9720 waiting at the head of an empty eastbound grain train as CP 9738 (with 8912 out of the picture) departs from Golden to Calgary with a mixed freight.

This is a member of another class of rebuilt locomotives; 5007 is an SD30C-ECO, a 3,000hp six-axle locomotive, one of a class of 50 that CP had rebuilt from former SD40-2 locomotives. 5007 was rebuilt from CP 6056 and although the frames and trucks were kept, the SD30C-ECOs received a new cab, electrics, prime mover and fuel tank. 5007 was photographed between duties at Golden on 26 May 2017.

A track evaluation train arrives at Golden on 26 May 2017 being pushed from the rear by CP 2241.

After a brief stop, 2241 pushes the track evaluation train towards KC Junction at the west end of Golden on 26 May 2017.

A pair of Canadian Pacific SD40-2s, 5757 and 6610, had just brought a train from Cranbrook into Golden and had left a set of loaded cars for onward movement to Vancouver in a siding on 5 August 2006.

A rather weathered 5730, an SD40-2, was about to work a local trip freight and was shunting to collect the caboose seen in the distance; part of the return journey to Golden would involve 5730 pushing its train with the shunter riding in the caboose and being in contact with the engineer by radio.

After pushing the track evaluation train to just west of KC Junction, the crew changed ends and now with 2241 leading, the train is about to travel south along the Windermere Sub to Fort Steele before it heads east through Crowsnest Pass to Lethbridge.

The line from KC Junction that 2241 took takes trains to Golden yard and the heavy repair workshops, and shortly after leaving KC Junction it crosses the Kicking Horse River. Three fairly new ES44AC locomotives are hauling coal empties returning from Vancouver to Sparwood and, with 8815 leading 8895 and 8807, they are taking their train south on 3 October 2011.

RM 8017 and 8012 pass through Golden on 4 August 2006 with the Rocky Mountaineer on its way to Banff.

I was able to photograph one of the last remaining Canadian Pacific SW1200RSm Class locomotives, along with a caboose, at the works at Golden on 16 October 2007. Canadian Pacific sold 1245 to the Ontario Southland Railway in October 2012.

SD40-2 5960 sits outside the workshops at Golden on 2 October 2011 waiting for repair.

An eastbound intermodal arrives at Golden on the evening of 31 July 2006 on its way to Calgary behind CP 8617 and 8575.

Canadian Pacific 9661 and 9720 are waiting at KC Junction, Golden, on 26 May 2017 for a loaded coal train to come off the line from Fort Steele; once it has gone, the train will cross over and head off towards Calgary. In the distance are the Selkirk Mountains, which the train has gone through on Rogers Pass on its way from Revelstoke to Golden.

The coal train has passed and 9661 and 9720 are now on the move with their eastbound grain empties, passing over the crossing from where I had taken the previous photograph on 26 May 2017.

In what is often called the 'blue hour' after sunset, an eastbound potash waits at Golden on 2 October 2011 for a westbound to come off the single line through the Kicking Horse Canyon. 8655 was the locomotive on the rear of the empty potash, which had its wagons illuminated by the street lights and the headlights of the vehicles passing along the adjacent road.

Chapter 6
Cowley to Blairmore

Not everyone knows that as well as the main line over Kicking Horse Pass, the Canadian Pacific Railway also has a second line through the Rockies running close to the US border in Southern Alberta and British Columbia, and crossing the Continental Divide at Crowsnest Pass. The line runs west from Lethbridge and after running through the pass, it arrives at Fort Steele where a line branches off and runs north to Golden while the original line continues on to Cranbrook. I will only cover the part of the line from Cowley on the High Prairies to the outskirts of Sparwood on the western end of the actual pass.

The Crowsnest line was constructed in only 18 months and was built principally for two reasons. The first was that numerous minerals had been found in southern British Columbia and vast deposits of coal had been discovered in the Elk Valley to the north of Sparwood in British Columbia and in Crowsnest Pass on the Alberta side. The second reason was that in Northern Montana, the US Great Northern Railroad was planning to build branch lines into southern Canada to tap into this potential traffic and the Canadian Pacific Railway received Federal Funding to help construct the line and stop a potential movement for communities in the south of Alberta wanting to join Montana and become part of the USA. The railway was built with these aims and it was always going to be a secondary line to the main line from Calgary to Vancouver via Kicking Horse and Rogers Pass.

After Cranbrook, a line continues south to the US border and trains use this line, the former Spokane International line, to run to Spokane, Washington and beyond. North of the border it is a CP line, south of the border it becomes a Union Pacific line and this will explain why so many of the trains that I have photographed over the years on the Crowsnest Sub have yellow Union Pacific locomotives as well as CP red ones.

Cowley is a small village with a great view of the Rockies to the west and a passing siding on the railway. Its main claim to fame was when it was used in the 2006 film Brokeback Mountain as the fictional town of Signal, Wyoming. Cowley only has a population of just over 200 and is at a height of 3,855ft above sea level but as the line runs westwards, it actually descends past the village of Lundbreck and keeps reducing in height until it crosses the Crowsnest River at the scenic location of Lundbreck Falls. From this point, there are a few further undulating parts of the line as it passes the siding at Burmis and runs towards the Frank Slide, the major feature on both the railway and the adjacent Highway 3.

The Frank Slide is the site of Canada's worst rockslide. In the early hours of 29 April 1903, the top centre of Turtle Mountain collapsed and 90 million tons of limestone fell, completely destroying the eastern half of the town of Frank, which was situated at the foot of the mountain and it left a debris field that spreads out for over two miles from the foot of the mountain. There was a coal mine in Turtle Mountain, which is now known to have an unstable geological make up. A series of large and deep cracks in the rock at the summit had filled with water that had frozen and it is thought that this plus vibration from the mine could also have been a significant contribution to causing the disaster. A locomotive had been switching loaded coal cars at the mine siding and just managed to get clear before everything was swept away. After the fall had stopped, a brakeman from the freight train managed to cross the fallen rocks and stop a westbound passenger train just before it ran into the rockslide. Today, loaded heavy trains have to pass below the mountain at a slow speed so as to minimise vibration and lower the risk of any further rock falls.

After passing Frank, the line curves round the shoulder of Turtle Mountain and arrives at Blairmore, the main town in the Municipality of Crowsnest with a population of just over 2,000 and at an altitude of 4,000ft above sea level. Displayed on a short length of track surrounded by a security fence to the south of the main street sits a 2-6-0 steam locomotive on display. Although the locomotive bears a No.1 on its smokebox door, it is actually the former Hillcrest Collieries No,11, built by the Canadian Locomotive Company at Kingston, Ontario in May 1914. It spent all of its working life at local collieries and after the Hillcrest mines closed, it moved to the Greenhill Mine and worked there until it was retired in 1961 when a group of local businessmen purchased it and put it on display in Blairmore.

Left: 6080, an SD40-2, is in the siding at Cowley with a track evaluation train that was going to Lethbridge on 30 September 2011.

Below: Union Pacific SD9043MACs 8290 and 8264 are slowly negotiating the reverse curves between Lundbreck and Lundbreck Falls on 4 June 2009.

After several days of heavy rain, the Crowsnest River was in spate on 17 June 2010 as UP 8264 and CP 9508 crossed at Lundbreck Falls with potash empties to Lethbridge.

Although it was only the first week of October in 2018, wintry conditions had arrived in Southern Alberta with a good covering of snow even down to lower levels. On 6 October, a very colourful combination of CP 9730, Kansas City Southern 4620 and UP 7802 approached Burmis with a loaded grain train that would travel to Cranbrook and then, by traversing the Spokane International line, to Spokane, Washington.

After collecting some loaded wagons from the siding at Burmis, UP 8269 and CP 9578 restart their train, which was heading for the yard at Lethbridge on 30 September 2011.

The rear locomotive on a grain train heading to Spokane, Washington, on 12 October 2012 was CP 8939, photographed near Hillcrest with Turtle Mountain and the Frank Slide as the backdrop.

There is normally an excellent view from this location of the Frank Slide but about ten minutes before the train arrived so did the clouds and an absolute deluge followed. Union Pacific 8264 and Canadian Pacific 9508 are heading to the yard at Lethbridge with an empty potash train.

The scale of the disaster called the Frank Slide is clearly seen as 8939 on the rear of a westbound loaded grain approached the slide on 12 October 2010.

A pair of Union Pacific SD9043MACs with 8265 leading 8284 start their run on 17 June 2010 through the debris field caused by the Frank Slide with a heavy loaded grain train. Heavy trains are required to pass through the slide debris field at a greatly reduced speed as the risk of a further fall does exist and running at slow speed lowers the vibrations the trains cause and also the risk of any accidents.

On an extremely wet 17 June 2010, a mixed freight from Lethbridge to Cranbrook approaches Blairmore behind a pair of ES44ACs, with CP 8736 leading 8716.

Union Pacific 8272 and 8263 take a westbound loaded grain through Blairmore on 12 October 2012.

Blairmore has a steam locomotive on display that at one time worked for local coal mining companies. Although it carries the No.1, this was at one time Hillcrest Collieries No.11, a 2-6-0 built in May 1914 by the Canadian Locomotive Company at Kingston, Ontario. After the Hillcrest mines closed in 1939, it worked at Greenhill until it was retired in 1961 but instead of being sold for scrap, the locomotive was purchased by a group of local businessmen and was eventually moved to its present location in 1967.

Another UP/CP combination on 23 September 2013 passed Blairmore, this time with UP 5529 leading CP 9591 on a mixed consist freight from Lethbridge to Spokane, Washington.

Canadian Pacific 9730, Kansas City Southern 4620 and Union Pacific 7802 haul a loaded grain past Blairmore on 6 October 2018 and are about to start the stiff climb to the summit of Crowsnest Pass.

The loaded grain that had CP 9730 on the front had CP 8026 as distributed power on the rear. 8026 was formerly AC4400CW 9581 but was rebuilt to become an AC4400CWM in 2017.

A pair of Union Pacific GE ES44ACs, 5523 and 5548, both built in the spring of 2005 are in charge of this mixed freight on its way to Lethbridge, photographed as it approaches Blairmore on 23 September 2013.

Chapter 7

Blairmore to Sparwood

As the line leaves Blairmore, there is a passing siding that ends at Coleman, the next community in the pass. Coleman has a population of more than 1,000 and at one time it was the largest and most important place in the pass. Its colliery's most memorable features were coke ovens, more than 200 of them were in use in the early 20th century. Like the rest of the collieries on the Alberta side of the pass, it closed towards the end of the last century as on the Alberta side, the coal was deep mined whereas on the British Columbia side it was opencast mining, which was much more profitable. The mines in the pass were also the scenes of some of Canada's worst mining accidents. At 9.30am on 19 June 1914, a terrible explosion swept through the Hillcrest mine, killing 189 out of the total of 228 who were underground at that time, making it Canada's worst ever mining disaster.

At the west end of the siding, the climb to Crowsnest Summit really gets under way with short parts of the line having a gradient steeper than 1 in 50. After a short distance, the line reaches Sentinel where there is a siding that serves the adjacent works. From Sentinel, the line continues now on an easier gradient, passing under Highway 3 (a good photographic location) until it levels out and runs on a ledge cut out of the mountain alongside Crowsnest Lake. This has been the scene of several accidents over the years, the latest being as recently as early March 2021 when some 20 loaded potash wagons derailed, some ending up in the lake, which made the task of reopening the line more difficult. After passing the west end of Crowsnest Lake, the line curves to the south and runs once more under Highway 3 before entering the passing sidings at Crowsnest Summit, which is often a crew-change point that has a Canadian Pacific crew accommodation building.

The boundary between Alberta and British Columbia straddles the line here and once the west end of the siding is reached, the line starts to descend towards Sparwood. At McGillivray, it runs in almost a 360 degree curve to lose height and then crosses under Highway 3. Until 2010, the line crossed Highway 3 on the level but a new road alignment with a bridge over the adjacent river eliminated a location that had seen numerous minor accidents occur between road and rail traffic. The line crosses Crowsnest Creek and curves sharply round before running alongside the north bank of the creek until it reaches Sparwood, where it makes a junction with the line from the Elk Valley coal mines before striking westwards towards Fernie. The approach to Sparwood is where this journey over Crowsnest Pass will end.

Two of Canadian Pacific's AC4400CWs, 8551 and 9586, pass the west end of Chinook Siding at Coleman on 13 October 2012 with a mixed westbound freight from Lethbridge to Cranbrook.

CP 8939 is on the rear of a westbound grain and is pushing hard as the train slowly grinds its way up the pass towards Sentinel on 12 October 2012.

CP red 9752 and CEFX blue 1032 are on the front of a heavy westbound grain that was slowly toiling up the steep grade at Sentinel to the summit of Crowsnest Pass at 4,455ft above sea level.

Pushing hard on the rear of the grain train at Sentinel was Union Pacific SD9043MAC 8297, photographed in a lucky burst of sunshine but under a stormy sky. Crowsnest Mountain at 9,137ft in height is in the distance.

A colourful combination was on the front of this westbound mixed freight passing Sentinel on 28 September 2018, CP 8560, a 1998-built GE AC4400CW, and KCS (Kansas City Southern) 4145, an EMD SD70Ace that had been built in 2014. In all my previous visits to the pass, I had never photographed a single KCS locomotive before, yet on this day I photographed four on four different trains!

A pair of SD40-2s in an obsolete CP red livery approach the Highway 3 overbridge at Sentinel near the summit of Crowsnest Pass with a rake of empty stone hoppers on 12 October 2012. The train was going to Cranbrook to be loaded; the stone was being used in the construction of the new runway at Calgary International Airport that was being built at that time.

Taken from the Highway 3 overbridge, a pair of Union Pacific ES44ACs, 5523 leading 5548, are about to start the descent on the eastern side of the pass with a mixed freight on its way to Lethbridge on 23 September 2013.

CP 9717 and 8507 are at the summit of Crowsnest Pass on 1 June 2017 with an eastbound empty grain train on its way to Lethbridge that had originated in Oregon, USA.

The empty stone train behind SD40-2s 5974 and 5912 on 12 October 2012 looks like a toy train as it runs along a shelf cut out of the mountain on the north side of Crowsnest Lake. The lake has a maximum depth of 89ft and an average depth of 45ft. The line at this point was built between 1897 and 1898 and rail traffic always has to be vigilant for falling rocks and slides.

Another SD40-2, 6080 sits at Crowsnest Summit Siding with a track evaluation test train on 30 September 2011 that was going to Lethbridge. I was able to photograph it and have a good look at it as it was waiting for a crew to arrive by a road transfer vehicle.

The CP/KCS westbound mixed freight with CP 8560 and KCS 4145 has arrived at Crowsnest Summit Siding on 28 September 2018 and has stopped for a crew change before it will continue its journey west to Cranbrook.

Union Pacific 5523 and 5548 start an eastbound mixed freight to Lethbridge out of the siding at Crowsnest Summit on 23 September 2013 where they had been held to let a westbound loaded potash go past.

If a crew are out of hours, it is common to find that a train will be tied down and left waiting until its replacement crew arrives, which considering the distances by road can be quite some time. On 21 June 2010, CP 8852 and 9598 are tied down at the west end of the siding at Crowsnest Summit with a westbound sulphur train.

Union Pacific 5529 and CP 9581 were also waiting for a fresh crew to arrive from Sparwood on 23 September 2013. Crowsnest Pass is rich in minerals, especially coal, but I did wonder if maybe there was also the proverbial pot of gold somewhere near to the other end of the summit siding!

When trains leave Sparwood heading for the short but stiff climb up to the summit of Crowsnest Pass, the line initially follows the Crowsnest Creek until at McGillivray it turns sharply to the south, crosses the creek seen in the foreground, passes under Highway 3 and climbs in almost a full circle in order to gain height. CP 8824 and 8787 were hauling a lengthy mixed freight to Lethbridge on 5 June 2009.

Just on the eastern outskirts of Sparwood BC, a long train of empty potash hoppers has five Union Pacific SD9043MACs on the front on 20 June 2010, in order from the front 8306, 8268, 8260, 8289 and 8277, which were waiting for a westbound to arrive off the single line after descending the west side of Crowsnest Pass. The train only required two locomotives, the other three were what is called a 'power move', taking locomotives to a place where they are needed to work a train, in this case the yard at Lethbridge.

Further reading from

As Europe's leading transport publisher, we produce a wide range of railway magazines and bookazines.

Visit: shop.keypublishing.com for more details